Charlotte Rosalys Jones

The Hypnotic Experiment of Dr. Reeves

And Other Stories

Charlotte Rosalys Jones

The Hypnotic Experiment of Dr. Reeves
And Other Stories

ISBN/EAN: 9783337004699

Printed in Europe, USA, Canada, Australia, Japan

Cover: Foto ©berggeist007 / pixelio.de

More available books at **www.hansebooks.com**

THE

Hypnotic Experiment

OF

Dr. Reeves

And other Stories

BY

CHARLOTTE ROSALYS JONES

London

BLISS, SANDS AND FOSTER

CRAVEN STREET, STRAND

1894

I

The Hypnotic Experiment of Dr. Reeves.

D R. EDWARD REEVES, the celebrated Rheumatism Specialist,
is not a favourite with the members of
his profession.   His methods of treatment being unknown, coupled with
his refusal as yet to divulge them,
have given his enemies and rivals a
chance to accuse him of charlatanism;
but to the great rheumatic public he
has become a demi-god; and as long
as our changeable climate continues to
nurture this disease, his idiosyncrasies
will be overlooked by the multitudes
whom he relieves.

In his genial moods, the doctor tells many curious anecdotes, and how some of his daring experiments were made under rather romantic circumstances. One of the strangest of them can best be told in his own language :

"Some time ago, I had, among my patients, a young man who interested me from the first. He came to my private hospital for treatment of a severe form of rheumatism of the heart; he was attended by a younger brother, whose devotion struck me as remarkable, until I became better acquainted with the invalid, and discovered how worthy he was of it all. He seldom spoke of himself, except his one great desire to get rid of the subtle disease that overshadowed his life, and he seemed anxious to aid me in every way with the treatment. Evidently wealthy, gifted, and just about eight-and-twenty, it seemed almost impossible to believe his bright young life was constantly threatened by the convulsive attacks which had become more and more frequent.

"Unlike most of my patients afflicted by the same trouble, he did not respond to the usual remedies; and I realized that if his life were saved at all it must be by employing heroic measures. However, sure that the disease was lessening its hold in general, and only needed driving away from a vital point, I awaited developments.

"Late one evening, as I was seated in my study, puzzling my brain with some questions of hypnotic influence over patients at critical moments, my night bell rang. I went to the door myself, and found there the nurse of my young friend, who told me my presence was desired at once, as the most alarming symptoms had reappeared. Stepping back for my hat, my eyes fell upon the book of *Experiments in Hypnotism*, which an old Professor in Paris had sent me, remembering my absorbing interest in Charcot's specialty, and a certain power I had developed when a student in the Latin quarter. This power I had used to tranquillise nervous patients, or to play practical jokes on my friends, after the manner of most young medical students who discover they have any skill in this direction. An idea occurred to me— Why not inoculate my patient with the powerful amount of virus required to drive the disease finally from the dangerous region of the heart, *while he is in a hypnotic condition?*

"In an instant after, all the perils of the situation presented themselves: Do remedies act if the patient is under this influence? Will the final result be the desired one? Providing the pain be temporarily stilled, would it re-occur after the hypnotic influence had been removed?

"These and other doubts so disturbed me, that on my way to the hospital I determined to avoid taking any such measures, unless I found the patient actually dying.

"As I entered, I was met by the brother. He seemed plunged in despair.

"'He is going fast, doctor,' he said. 'Can you do nothing?'

"Without a word I stepped to the bedside. I found my worst fears realized. At a glance I saw he would not survive the night unless the frightful spasms that were fast sapping his strength were arrested.

"As I took his hand and felt his pulse, he looked up past me at his brother, and gasped the one word 'Annie.'

"'Whom does he want?' I asked.

"'His *fiancée*, doctor. My brother

was to have been married in a month; but when he knew that he was threatened with a probably fatal disease, he begged me to help him quite secretly to try this last chance for recovery; and so, although he is within a mile of his own house and that of his intended wife, no one but myself and his faithful servant is aware of it. To all our friends we are hundreds of miles away, looking after business interests. And now it has grown worse and worse, until he is dying, absolutely within reach of Annie, to whom he is madly devoted.'

"'Will you be calm, and help me to make one last great trial?' I asked.

"'Great heavens! What can I do?' he replied.

"'Take my carriage; it is at the door; tell the coachman to drive his fastest to Annie's house. Bring her back with you; and, above all, explain to her the situation, so that I can count on perfect calmness."

"Without a word he was gone, and as I heard the wheels leaving the door, I turned back to collect my thoughts for a moment before returning to the sick-room. I had to count on at least half-an-hour's delay, and meanwhile to quiet this horrible pain and wait for Annie to help me.

"Once back in my patient's presence, I took his hand, looked fixedly at him until his eyes caught mine. Then I said, 'You must sleep now; Annie is coming, and you must be strong to see her.'

"At once a look of surprise, of joy, followed by one of despair, passed over his face. 'I am dying, and you have sent for her,' he murmured.

"'Sleep,' I said, this time completely fixing his gaze. Almost instantly the spasms ceased, and he sank back among his pillows like a tired child.

Not noticing the look of astonishment in the face of the nurse (who was a faithful old valet of the invalid), I ordered him to send me the assistant-surgeon and a bright young woman nurse, whom I often selected for urgent cases. They came at once. It was the work of a few moments to inoculate the greatest quantity of the powerful poison that I had ever used at any one time. I then made the usual passes, and awoke the patient, resolved not to risk any unnecessary complications. I knew if his strength could be kept up for three, or at the most four hours, the battle was ours. But could he fight it out alone? I did not dare to guarantee the usual result of the virus if he were asleep. I could only count on Annie's support to help him out, for he seemed at last ready to give up the fight. Even now the impression that his sweetheart was coming, added to the rest secured by the little respite from pain, seemed to be sustaining him, and all I dreaded was, that he would be too feeble to bear the effects of the remedy in its later processes, when the convulsive attacks were liable to be especially violent, as if they knew they were losing their power over their victim.

"A half hour passed, then three quarters, and I heard the wheels stop outside. I opened the door, went softly into the hall, and met the brother, pale, anxious, and—alone !

"'She is not at home, doctor. She is at a ball, believing my brother well and hundreds of miles away. I explained all to her father. He has gone to fetch her. Am I too late?'

"Just then a moan from the adjoining room told that my patient was suffering. I returned quickly to his bedside, and found the old symptoms reviving. Again the temptation beset me. I argued: 'I influenced him easily, he certainly feels no pain while hypnotised, he cannot live unaided through another convulsive attack. To be sure, I have to fear that he can never be awakened, and that the final effects of the remedy may be lessened. At least two hours must elapse before he is safe, providing no new complications set in; and meanwhile what an opportunity to see if hypnotism prevents or aids inoculation! He has no other chance. The plan of fighting it out on natural lines, aided by his own desire to live for his love's sake, has failed.'

"I hesitated no longer. Again taking his hand, I uttered the magic word 'sleep,' and he sank back as before.

"'Now for the great *coup*,' I said, and, turning to my young nurse, I ordered her to take off her cap, put on a hat and cloak, and follow exactly the few directions I gave her. She seemed to grasp my idea, and left me free to follow out my experiment.

"'Annie is coming,' I said, looking straight into the poor fellow's eyes. 'In a few minutes she will be here.' I hesitated even as I spoke. Can a hypnotised patient be made to believe that a *substituted* person is the one he expects to see? But even as the thought flashed across my mind the door opened and the brother entered, with the young nurse on his arm dressed in walking costume.

"'Here is Annie,' I said. There was a moment of horrible suspense. Then at a sign from me the young woman approached the bed, sank down on her knees, and took both his hands in hers. A look of incredulity, of wonder, of hope, and then one of ineffable peace shone on his face.

"'Annie, dearest, I tried to keep it from you and to come back to you free from this terrible trouble,' he whispered.

"'Yes, dear. You must not talk to

me. See, I am with you, I shall not stir.'

"'Kiss me once,' he murmured.

"The woman reached gently over and kissed him, and with his hands still in hers he relapsed into unconsciousness. In an hour more the danger would be over, but I must then awaken him, and unless the real Annie were present the shock might ruin everything. The moments went by —the sick man sleeping, the tireless nurse kneeling in her strained position by his bed, the brother pacing up and down outside the door, and I, watch in hand, dreading the last act in this exciting night's drama.

"Fifteen minutes more and I heard a rustle, a murmur of voices broken by sobs, and then silence. Suddenly and quickly the door opened; a beautiful woman in ball costume, with jewels gleaming in her hair and on her neck, glided like a spirit to the bedside. The nurse, with a woman's quick intuition, softly withdrew her hands; the other knelt and took her place, and with her eyes fixed on the face of him whom she had thought far away and in perfect health, she waited. 'She is worthy of him,' I thought, as I saw her attitude and her wonderful self-possession.

"Now for the test. Motioning the others to leave the room, I awoke the sleeper. His gaze instantly fell upon his Annie's loved face. 'Hush!' she said; 'you have been very ill; I know all about it; but the danger is over; all will be well, and I shall not leave you.' A puzzled look swept over his countenance. Then he feebly whispered, 'I was dreaming you were here; but you had your hat on, dear; you had just come in from the street and found me.'

"'I am *really* here,' Annie replied, 'and you must reward me by not saying another word.' She smiled at him, a brave smile, through the tears that were coming now. This time, with a satisfied look, he fell into a natural sleep. I knew the danger was over, and that I could safely leave him with his own.

"As I passed out into the early misty morning I confess the thought of the success of my part of the experiment was rather swallowed up by my admiration for that woman, and for the love, the great unselfish, protecting love, she had won from that man. Visions of lost happiness came before me, and it seemed to me I had missed something which might have been mine, had I been less absorbed in other ways; but just then I reached my own door, and caught sight of the name on the small silver plate, *Dr. Edward Reeves.* And I thought of the material I had collected for a medical paper on that night's work, so I dropped the sentiment, and

went in to make a record of the facts
in the case (which has interested the
scientific world ever since) of a patient
actually getting the full benefit of a
remedy while in an undisturbed, hypno-
tized state, despite all theories to the
contrary."

II

An International Courtsbip.

IN seven minutes the great steamer
Lahn would slip her moorings and
sail for Southampton.

Already the more cautious friends
of the departing passengers had left
the ship, and were finding places on
the dock whence they might wave
their final messages. The decks were
clearing fast, leaving mournful groups
of travellers, who were beginning to
realize how soon they would form a
little world of their own ; and so they
were making quiet observations of each
other.

A tall, sturdy, young Englishman
was leaning over the rail, looking a
trifle amused at the scene about him,
and occasionally waving his hand to
two men on the wharf, who were
evidently "seeing him off." He did
not look particularly sad, or as if he
had any especial interest in the voyage
beyond reaching his destination. That
he was distinctly a well-bred Englishman, who knew his London well, one
could not doubt; that he was also a
trifle obstinate, might be surmised
from the pose of the intellectual head
upon the square shoulders, and the
determined look about the firm, well-
shaped mouth. Just now, he has
screwed an eye-glass into his eye, and
is looking at two ladies who have
crossed the plank, and are being
greeted by two elderly gentlemen,
each of whom presents them with
bunches of flowers.

Something about them strikes the

young man's fancy; perhaps he is
interested in seeing that they seem
quite oblivious of the fact that the
warning bell is ringing, and he is wondering if the two men are to sail also,
when suddenly, just as the gangway is
to be removed, he sees them all shake
hands, and the two women are left
standing alone.

After a final look at his friends on
the dock, he takes a turn about the
steamer, and far off on the side, quite
removed from the harbour, he sees
the younger woman standing, looking
out—not behind, at what she is leaving,
but before her. *Why* it is that he
cares at all what a perfectly unknown
young woman is doing or thinking
puzzles Mr. Gordon-Treherne. In his
five-and-thirty years, he has known a
great many of the fair sex; he has
had several rather close love affairs—
with various results. He was rescued
from what might have terminated in

an unfortunate marriage when in Cambridge. The Gordon-Trehernes considered that the heir of the family had no right to throw himself away upon a modest little English girl, even if she were the daughter of the rector, and deeply in love with the fascinating young collegian.

After that experience, young Mr. Gordon-Treherne, or "The Arab," as his chums called him, from his love of travel, determined not to hurry himself about marrying. One or two charming Frenchwomen almost destroyed this resolution, and once he was decidedly fascinated with the daughter of an English general out in India. But he had travelled the length and breadth of the United States, and never felt inclined to fall in love with an American girl. Several of his friends had married American belles; and when young Lord Clanmore's engagement was announced to the beautiful and wealthy Miss Lawson, of New York, everyone envied him ; but Treherne had not cared to enter the lists, although he knew Miss Lawson well. Women said he was a man with a history, but he was all the more fascinating for that. Men called him a good fellow, and said "The Arab" was the best shot and the coolest rider in the club, only he was always running off to some outlandish place, where his accomplishments were lost.

Just now his friends might have been surprised to see him arranging a steamer chair for the elder of the two women who had caught his attention on the dock. The steamer has left the quay only a half hour, and already an opportunity has presented itself to make their acqaintance. Etiquette at sea is very elastic, and it only needed the usual attentions to the comfort of the elder woman to

attract the notice of the younger. She
has turned now, and with her hands
still full of flowers, comes toward them
—a tall, slim girl, possibly four-and-
twenty he thinks. He is dimly con-
scious that both ladies are quietly
but elegantly dressed. Americans, he
fancies; and then the elder woman
speaks,—"Thanks, so much." The
voice is low and musical. She must
be French, he thinks. She is a
brunette, and he decides that she
cannot be the mother of the tall,
fair girl who seats herself next to
her.

"Let me arrange your rug also,"
Mr. Gordon-Treherne says, as he
raises his hat.

"Oh, thank you; that is very comfortable."

And again he is struck with the
well modulated tones, which he
scarcely associates with American
voices.

Still they must be Americans, the young man argues to himself, but no longer finding an excuse to tarry in their vicinity, he moves off, and they meet no more till dinner-time.

Meanwhile, with the philosophy of an old traveller, Mr. Gordon-Treherne has interviewed the head steward, and, foregoing the honour of sitting at the captain's table, he has asked to be placed at a small one with a sofa-seat. Experience during previous voyages has taught him that there are certain comforts not to be despised in a side seat under a strong light. He sees several prospective lonely evenings, when he may not feel inclined to hunt about for a good place to read.

At dinner Mr. Gordon-Treherne notices two elderly men and a small boy at his table, and remarks two vacant places. Presently his two interesting acquaintances of the morning appear, and he has just time to read the cards on the plates on either side of him—"Mrs. Barry" on one, and "Miss Stuyvesant" on the other —and to comprehend that by some blunder he is separating them, and that he can only remedy the matter by giving up his cherished seat, when the two ladies arrive at their places. There is a moment's hesitation, and Mr. Gordon-Treherne remarks, "Allow me to change my place." Suiting the action to the word, he steps past and allows Mrs. Barry to take his seat, which brings him opposite Miss Stuyvesant. Both ladies express their thanks, and then, naturally, they fall into conversation. They speak of the steamers; Mr. Gordon-Treherne prefers a larger boat, and refers to several "ocean greyhounds" he has personally known. Curiously enough the ladies have made the same crossings, but prefer even smaller steamers than the

Lahn. "Americans, surely; 'Globe Trotters,'" he thinks.

He mentions that he has just been to the Exhibition at Chicago. Miss Stuyvesant says that in point of exhibits she preferred the Paris Exposition of '89, and so on, until it seems as if there were no place this young woman had not seen and about which she had not formed her conclusions. He doesn't care for it though, Arab that he is; he likes to travel, but the women of his family have never expressed a desire to go beyond Paris, and he thinks promiscuous sight-seeing outside a woman's province. He shows a little of this in his manner, for as he leaves the table, the elder woman says:

"How glad I am, Helen, that you do not believe in International marriages. Now here is a well-bred, intelligent Englishman, yet he shows insensibly what narrow ideas he has about women. I admit he is polite, and careful in small details of manner, but an American girl of spirit could never be happily married to him. Their ideas of life are too opposed."

Miss Stuyvesant has evidently not thought much about him, for she only smiles in a vague way, and says she has learned not to quarrel with the old-fashioned notions of English people.

"Why, I pride myself in actually leading them, when they start in a tirade against the very things I do myself!" she said.

"You are a sadly worldly young woman," Mrs. Barry rejoins, "and I wish you would marry and settle in your own country."

Meanwhile Mr. Gordon-Treherne was idly pacing the deck, smoking his cigar, and wondering if the self-possessed young woman would appear

later on. "If ever I marry," he re-
solves, "it will be to some woman
who has *not* been everywhere and
seen everything. I should feel as if
I were travelling with an animated
guide-book. I wonder if that girl has
a home?"

Then it occurred to him that Miss
Stuyvesant had merely answered his
questions, and as these had been
restricted to quite impersonal topics,
he only knew her name after all.

That she was good-looking, agreeable, and witty, he had already observed,
but she did not seem to thrust any
information about herself upon him,
as he had supposed an American girl
would. He did not see her again
that day, nor till the next afternoon,
when she was walking up and down
the deck with the captain of the
steamer, and as she passed him with a
little nod of recognition he heard her
speaking German.

"Surely American," he thinks, "knows the captain already, and speaks his language."

At dinner Mrs. Barry was missing, but Miss Stuyvesant appeared looking as calm and "well-groomed" as if a heavy sea were not tossing everything about, and obliging the passengers to eat over racks.

"You are an old sailor, I see," began Mr. Gordon-Treherne, "but I fear Mrs. Barry is ill."

"Yes, quite seriously ill," Miss Stuyvesant replied. "It is always an ordeal for her to cross the ocean."

"And has she done so frequently?" he asked.

"Nine times with me," the young woman coolly replied.

"Really," he said with a smile, "one might infer you had some designs on her life, did you not look so anxious about her."

"Oh, no, we usually have some

excellent reason, we do not take this voyage in order to martyrize Mrs. Barry," she replied.

"I shall have to ask her nationality outright," he thought.

"Then you do not live in America all the time?" he said.

"Not now, we are 'birds of passage,' and, like them, follow the spring-time; our habitation is usually settled by the climate."

"And do you know England?" he asked.

"Quite well, I was at school in England, and some of my dearest friends are living there."

"Some church school," he mentally remarked.

"Ah, then, perhaps you do not altogether despise our little island, and look down upon us from your bigness with the scorn that most of your compatriots do?"

"He is trying to make sport. I

shall foil him," she thought, and quite calmly said—

"Look down upon a country upon whose possessions 'the sun never sets'? Besides, the fact that I stay so much in England ought to prove how much I admire most of its institutions."

"Clever girl!" he thought, "trying to be a little satirical, and doesn't commit herself as to *all* of our 'institutions.' I must make her angry to get her real opinion."

And then he said, "You should see our English home-life. I am sure *that* appeals to every American woman."

There was a patronising tone about this remark that Mr. Gordon-Treherne felt would effect his purpose.

"Indeed," she said slowly, and went on eating, as if the conversation were beginning rather to bore her. Now, why Mr. Gordon-Treherne should

c

assume that Miss Stuyvesant had not seen this phase of England as well as others cannot be imagined; but there he overstepped the line, and soon after the decidedly cool "Indeed," Miss Stuyvesant left the table to look after her *chaperone.*

"An egotistical man," she thought, as she went to her state-room. She had liked Mr. Gordon-Treherne's appearance, and being a cosmopolitan young woman, was prepared to find him agreeable. Now she thinks him distinctly aggressive, with his old conservative ideas of women and English superiority.

He, for his part, feels he does not understand this American girl, who refused to quarrel with him, but suddenly turned and left him. He knows he has not shown himself in his most brilliant colours.

The days passed rapidly. Mr. Treherne and Miss Stuyvesant saw each other at table, walked the deck together, and to the casual observer seemed to be mutually entertained. But although they were in so many ways companionable, they both felt an intangible barrier between them in the national prejudice that their first conversation had developed—a prejudice probably latent in every person, however cultivated or travelled, although in this particular case both of these young people flattered themselves that they were singularly broad-minded.

The last evening of the voyage, as they were walking up and down the deck, Mr. Gordon-Treherne determined to broach the subject which he felt they had both avoided.

A larger acquaintance had brought out the fact that Miss Stuyvesant had read for honours at an English University, and Mr. Treherne was obliged to admit that in this case the higher education of women (which never strongly appealed to him), had not detracted from her personal charm. She, on the other hand, discovered that he knew a great deal about *her* country, and considered its possibilities almost unlimited; but she felt that he looked down upon its newness, and she resented his opinion of American men, whom he described as clever and agreeable in their relations with each other, but servile in their attitude toward women. The dangerous topic of national characteristics had not been touched upon until to-night.

Now Mr. Treherne is saying, " I hope you have forgiven my frankness in telling you exactly what my impressions were of America. I could not help seeing how charming and bright the women were, and I wondered if they did not despise the slavishness of their husbands and lovers. While the men are toiling to get rich, their families come abroad, their wives thus educating themselves beyond their husbands, and returning home, find themselves less than ever in sympathy with their surroundings. I never wonder when an American girl, who has had a chance to see the world, marries a foreigner of family and education."

If Mr. Treherne had been closely observing his companion, he might have remarked an ominous expression crossing her face, but she only said—

"I have had several friends in Europe whose fathers' fortunes have found them titles, and on the occasions that I have stayed with them, they did not seem wildly enthusiastic over the equality of companionship. The head of the house had generally gone to town, or was taking a run over to Paris, and I wondered if it suited a woman very well who had been accustomed to have a small court about her at home, to find herself restricted to a husband so little her companion that she scarcely ever saw him."

"But then you see, Miss Stuyvesant, she knows he is not down in Wall Street, or in some exchange, staking all his fortune on the rise and fall of stocks."

"No," she rejoined; "in the cases of my friends the women have to consider that their husbands are probably at Monte Carlo or Ostend. But really, why should we discuss it, Mr. Treherne? No one would ever fancy *you* admiring an American woman, and I, for my part, if I marry at all, would only marry an American man."

With which delightfully feminine declaration, Miss Stuyvesant says "Good-night," and abruptly leaves the astonished Treherne to realize that he has not made a good finish. Not that he cares seriously for Miss Stuyvesant; but Treherne is accustomed to find that women like him, and this girl, his instinct warns him, does *not* approve of him and his opinions. He feels annoyed, but there seems to be nothing to explain; his training and the circumstances of his life have made him conservative. He does not wish to love, nor does he especially approve of a young woman, however attractive, whose ideas differ from his own so materially.

And so next day, when he bids a formal "good-bye" to Mrs. Barry and Miss Stuyvesant, he tries to feel that in England he has more manly occupations than doing the agreeable to a young woman, and that woman an American. This is exactly what Mr. Treherne does *not* feel, nor does he mean to indicate it by his manner at parting. And so he goes off, consoling himself with the reflection that he certainly has found Miss Stuyvesant a pleasant companion for a sea voyage.

Three weeks later, in London, when
the season is at its height, Miss Stuyvesant, who is looking radiant in a
French gown, meets Mr. Gordon-
Treherne at Lady Clanmore's ball.
She is on the arm of the American
ambassador, and as she crosses the
room with that unconscious grace of
hers he feels that every man present
would be glad to know her, to talk
with her as he has talked, and something at that moment tells him that
she interests him more than any
woman has ever interested him before. Just then she sees him, and he
fancies that a rather annoyed look
crosses her face. Then she smiles,
and he comes over and speaks to her
and to her escort, who seems to know
everyone.

"Will you give me a dance, Miss
Stuyvesant?"

"Yes, but I have only this one
waltz left. You see, you English-

men *do* think that American girls are good partners—in a ball-room," she adds slyly.

" I see I am not forgiven," he says; and then the waltz begins.

What a waltz! Gordon-Treherne has had many good partners in his day, for he has always been a dancing man; but never has he seen anyone dance like this girl. When they stop she is scarcely out of breath, and he has only time to say, "Let me thank you." For her next partner had already claimed her, when she turned back and mischievously remarked, "And you, you dance extremely well —for an Englishman."

It occurred several times afterward to Miss Stuyvesant that he could do a great many things extremely well; and if he had only been born in America she might have preferred him to honest Jack Hamilton, who had loved her since she was a school girl, and who was doing exactly what Mr. Treherne had described in that last obnoxious conversation—staking his fortune in an Oil Exchange, hoping that some day he could induce Miss Stuyvesant to give up her Bohemian life for the luxuries of a wealthy American home. In an indefinite way she had thought she might do so in the end, but, while she gave no promise, she was sure that Jack would never change. And so she had drifted on pleasantly and thought-lessly, caring nothing for the men she met until this one, with his strong opinions, crossed her path, and for *him* she believed she entertained the most indifferent feelings. He had simply disturbed her. She did not think his ideas correct, but there was a sense of justice in the girl that made her think herself narrow and bigoted for not being able to judge things from other standpoints than her own. It was exactly what she was criticizing in Mr. Gordon-Treherne.

"It will be better to avoid any more discussions," she thought, and so the two did not meet again until one glorious autumn morning, when the house party at Lady Clanmore's rode out to the first meet of the season. Miss Stuyvesant headed the cavalcade, escorted by Lord Clanmore, and as they came up to the meet she saw Mr. Gordon-Treherne, who was riding a restive thoroughbred, and looking what he was—an excellent rider. He was talking to a handsome woman, beautifully gowned, who was driving a perfectly appointed trap.

"That is Lady Diana Gordon," Lord Clanmore is saying. "She is Treherne's cousin, and rumour has it that the old estates of Gordon and Treherne are liable to be joined."

Miss Stuyvesant feels for a moment as if she were slipping from her saddle, and then Treherne sees her. He raises his hat, and she smiles back an odd, unconsciously sad little smile, which he has only time to remark, when the hounds move off. And now all the recklessness in the girl is aroused; she knows she rides as few women can, and during the run she follows her pilot, Lord Clanmore, so straight that the whole field is lost in admiration of her.

Treherne alone has noticed the set look in her face. "Is she ill?" he wonders, and he determines to keep her well in view. He has hard work, for she is on a vicious little mare which she insisted upon riding, and as she takes fence after fence Treherne grows more and more anxious. The hounds have come to a check just beyond a clump of trees in the next field. Miss Stuyvesant turns her horse's head, and Treherne sees she intends to take a short cut through a dangerous low-

boughed copse which intervenes.
"Stop!" he calls, but she does not
hear him, and he knows his only plan
is to head her off, if possible. Turning
sharply, he enters the field from the
other side; as he does so, he hears
the crashing of boughs, and sees Miss
Stuyvesant's mare coming straight
towards him. Each moment he
expects to see her swept from her
saddle, but she keeps her seat bravely.
He calls out to her to turn to the right,
for before her in her present path is a
strong low-hanging branch of an old
oak, which Treherne knows she cannot
pass safely. An instant after, he sees
she has lost control over the mare, and
he heads his own horse straight towards
her. With a quick, skilful motion he
grasps her bridle just as the horses
meet. There is a mad plunge, and
Mr. Treherne, still clinging to the
other reins, has dropped his own and
is dragged from his saddle. He helps

the girl to dismount from her now subdued, but trembling animal. Miss Stuyvesant looks very white, and Mr. Treherne is offering her his hunting flask, when Lord Clanmore gallops back to them.

"Your empty saddle gave us a great scare, Treherne. Are you hurt?"

For Mr. Treherne, too, has suddenly grown very pale.

"It's nothing, Clanmore, just a little wrench I gave my arm; that's all."

And Miss Stuyvesant remembers how skilfully that arm had lifted her from her saddle. In that moment she knows she loves him. Every vestige of national prejudice is swept away, and poor Jack Hamilton's chances are gone for ever.

The next day Mr. Treherne managed to write a few words with his left hand and send them back by Miss

Stuyvesant's messenger, who came to enquire after him. He said—

"DEAR MISS STUYVESANT,

"Many thanks for your kind enquiries. I shall be restricted to using my left hand for a time, but I must tell you how plucky I thought you yesterday. The stupid doctor has forbidden me to leave the house, but unless you wish to increase my feverish symptoms please send me some token by this messenger to assure me you have forgotten my first impressions of your country. As soon as I am able I shall beg you in person to reconsider your decision about marrying 'only an American.' My happiness depends upon your marrying an Englishman who is

"Entirely yours,
"E. GORDON-TREHERNE."

When Miss Stuyvesant read this note she took two beautiful little silk flags—one a Union Jack, the other the Stars and Stripes, and tying them together with a lover's knot she sent them to Treherne.

*   *   *   *

In after years Mr. and Mrs. Gordon-Treherne's friends remark the deference which they pay to each other's ideas; and the entwined banners, which occupy a conspicuous place in the library, are called the "FLAGS OF TRUCE."

III.

One Woman's History out of Many.

"SISTER FAITHFUL" she was called by the Edgecombe people. Her name was really Faithful Farrington, but no one ever said "Miss Farrington." She had been born in the old Manor House, where for fifty years she had spent the most of her time. Her father, old Nathan Farrington, had been content to live the life of a recluse after his wife's death, finding his greatest happiness among his books, and in directing the education of his two children.. Francis Farrington, the son, had gone out to India in early life, and had risen high in the Civil Service. He had lost his

D

young wife, and after many years of valuable work had returned, an invalid, to Edgecombe, where he found in his sister the most tender and sympathetic of companions. He was content enough to allow the whole responsibility of the estate to rest upon her patient shoulders. She, for her part, grew up to know a great deal about science and literature, but absolutely nothing of society or the world. When she was thirty her father died, and, besides her brother out in India, and a distant cousin, who was a Professor in some London college, she had no one nearer than the old nurse who had tried to fill the place of a mother to her.

Having a considerable fortune, she lived on in her old home, attended by the same faithful servants, exactly as she had always done, except that when the long winter evenings became tedious, and books failed her, she invited some of her townswomen in to

tea and played a rubber of whist.
Her days were filled with good works;
every cottager in the neighbourhood
knew her, and she knew them and
sympathized with all their sorrows.
Wise in her charities, she was the
vicar's most invaluable assistant, and
it is to be doubted whether he, in his
rôle of spiritual adviser, was as much
loved and revered as "Sister Faithful,"
whose tireless hands constructed wonderful garments for the babies, and
whose name was borne by half the
children in consequence.

No breath from the outside world
had ever touched this woman. Once
she had gone to Paris with her father,
but it remained in her mind simply a
lovely picture, a little larger and more
daring in colour than the pictures she
had seen at the Louvre. She had
been up to London several times, but
that was to make notes at the British
Museum. Her life was in no way

different from that of most of the
women she had known.

Once she had seen an item in a
journal that struck her forcibly; it
mentioned that there were eight hun-
dred thousand more women than men
in Great Britain, and that a good
proportion of them were matrimonially
eligible women as regarded property
and accomplishments, but that they
were of the middle classes, where
marriage was most infrequent. Sister
Faithful had remarked then that she
knew a great many attractive women
who were not liable to marry. She
wondered why, for her education had
made her logical. And then she
reviewed her own life. All the male
members of the families of her friends
had gone to larger towns as soon as
they were old enough; the girls, after
a touch of boarding-school, had come
home to assume simple household
duties, and, an occasional curate ex-

cepted, they did not often meet young men.

"Sister Faithful," having for her constant companion a man who lived in books, had rather a better-trained mind than most women. It had not been allowed to wander, and her greatest weakness was her way of bestowing charity. She did not like to account to anyone for it, and so tired mothers, who sent their offspring to her for a holiday, were apt to have them returned in new and wholesome garments, which showed that a heart calculated by nature to be a motherly one was bestowing its bounty quietly on other women's children. Strange to say, Sister Faithful had not given any thought to marriage for herself. That she should ever leave her father, marry, and have children of her own seemed impossible. She was quite content to accept life as she found it, and improve the morals and manners of the children of the lower classes about her. And now she was fifty, and until her brother's return she had lived alone.

She had remembered yesterday that it was her birthday, and had celebrated it by inviting the school children to tea in her garden, which was in in its loveliest summer dress. In the evening she had received a letter from her distant and unknown cousin, the Professor, whom she had only met in those long ago excursions to London, saying he was "tired"—"worn out," his doctor said—and he had written to see if his cousin would take him in for a few weeks' vacation. "I shall live out-of-doors," he wrote, "and I promise in no way to disturb your life. I want only my books, and to wander about over your beautiful country." She wondered if she had been hasty when she wrote back to bid "welcome to our nearest of kin and our father's friend." She remembered that after all he was really a cousin once removed, and a little younger than herself, and that when her father had liked him he was very young indeed. A glance at the mirror re-assured her. She was very free from vanity, and she realized she was no longer young. The villagers called her beautiful, but perhaps their sensibilities, sharpened by the lack of beauty about them, were keener in detecting their benefactress' fine points.

Thanks to her healthy, regular life, Sister Faithful at fifty was very good to look upon. The soft hair, worn lightly back from the low, well-shaped forehead, was only faintly tinged with grey, and her skin was as smooth and fresh as that of a woman of half her age. It was not the firm, quiet mouth, nor even the gentle, sweet brown eyes that attracted one most; it was the unconsciousness of the woman, the very annihilation of self, as it were, without affectation, that made one long to know *why* she was constantly giving without any question of return. No man had ever told her she was beautiful; her father's friends had approved of her, but then she ministered to their comfort, when they came to stay at Edgecombe. She attended to everyone's wants, and seemed to have gone through life without dreaming that in some larger sphere she would have been considered a very attractive woman. Not that she was perfect; she had her idiosyncrasies—as who has not? but she had a disciplined, well-trained, unselfish nature, that overbalanced any faults. Even now her one consideration was for Cousin Emerson, who was to arrive the following day. Would he be comfortable in Edgecombe? Would he not be lonely with her and only an invalid host to look after him? These, and other doubts crossed her mind, and, as a relief, she spent the entire day overlooking the sweeping and dusting of the already clean house.

Next day, the evening train brought Cousin Emerson. As he alighted from the carriage, Sister Faithful thought him only an older edition of the intellectual-looking man she had met in London. He was evidently still ill, and looked as if he had burned too much midnight oil. Her practical mind immediately swept over the entire list of nourishing dishes that she might concoct for him. He, half-an-hour later, glanced over his well-appointed room, and thanked fortune that it had occurred to him to stay with his good cousins.

After dinner this occurred to him again as he stretched himself on the comfortable library lounge, and let the smoke of his cigar curl up in slow, bewitching rings about his head, while Cousin Faithful read aloud in that well-modulated voice of hers.

And so the days went on, bringing health and strength to Cousin Emerson, and great, unspeakable content to "Sister Faithful," as he too called her.

"Somehow," he said, after he had been at Edgecombe for several weeks, "it seems as if we were more than cousins. I shall reverse your name; you shall be my Faithful Sister, as you have been nurse and friend."

At first he had accompanied her in her long afternoon walks, when she visited her cottage people, but after a while he persuaded her to take all sorts of short excursions on foot, or again they would drive over the hills about the estate.

The evenings were perhaps the sweetest of all to Sister Faithful, for then her interests in the outer world ceased, and, until her cousin came, she had often felt very lonely. Now, they read aloud, played a friendly

game of cribbage, or strolled about
the garden when the nights were fine.
Autumn was drawing near. Cousin
Emerson's visit had lengthened to two
months, and still he said nothing
about going. He was quite strong
again, and seemed to have lost the
melancholy that at first overshadowed
him. Faithful's heart rejoiced as she
looked at him, and she did not allow
herself to think that it might end.

One morning, in early September,
the post brought several letters. They
were breakfasting. Faithful remembered every detail afterwards. The
pungent odour of chrysanthemums
always carried her back to that
morning. Cousin Emerson had
gathered for the breakfast-table the
splendid bunch that adorned it.

Suddenly a look of intense happiness
lighted up his face. " Faithful Sister,"
he said, looking across at her, " I want
you to be the first to congratulate me.

At last the woman whom I adore, for love of whom I have been so miserable, has consented to marry me. I doubt whether, if I had not fallen into your dear hospitable hands, I could have struggled so well to recover."

In his excitement, Cousin Emerson did not notice the pallor that swept over Faithful's face. Her voice was steady as she said, "Why have you not let us sympathize with you all along?"

"Oh, it all seemed so hopeless," he said, "and I could not bear to open the old wound; but I am to go up to London at once, and I shall bring my bride straight to Edgecombe, if I may."

That night he left, after many cordial expressions of gratitude, and Sister Faithful, apparently unmoved, saw him go; but afterward she had no mind to wander about the garden, or read a favourite book. She went

quietly to her room, and, for the first
time, wept.

She knew she had been companionable to this man; that in her
society he had found peace and
content. And yet—in a moment—he
had forgotten it all; he had gone to
win some other woman, impelled
by what he called love. Was it love
she felt for him? Even then, in her
loneliness, with a grey-skyed future
before her, and no prospect of change,
she felt only her own inconsistency.
"He was my kinsman and guest; he
never asked me to love him, and he
never knew my feeling for him," she
argued, and so the night passed, a
night of unselfish sorrow for the lonely
woman, while the man was being
whirled towards the one being who
engrossed *his* thoughts.

Afterward, when Cousin Emerson
and his wife came to Edgecombe to
visit, he remarked, in the privacy of

their room, that Cousin Faithful had aged terribly; but to the poor people she seemed more saintly than ever, for after that one happy summer — the only time she had ever allowed herself any personal happiness — she had returned to her charities as if she wished to make up for some neglect. And when the villagers called her "Sister Faithful," she felt it almost as a reproach that she had dared hope for any other name.

IV

Miss Cameron's Art Sale.

KATHERINE CAMERON was spending her third winter in Paris. The first year she had led a quiet, uneventful student's life. The second season she launched out a little into society as represented by the English and American colonies, and now she was spoken of as that "clever and rich Miss Cameron," whom the English-speaking residents remembered to have seen at various *musicales* the year before.

On her return from America, with the reputation of added wealth, she found herself invited everywhere. Everyone wondered that she did not marry, for she was a young woman whom men admired apart from her money and accomplishments. But although she went out a great deal, and was usually surrounded by a little court of struggling tenors and impecunious titles, she seemed unmoved by all the attention she received, and apparently was not even greatly amused.

The truth was, Katherine Cameron, being a clever girl, had seen through the artificiality of it all, and still could not bear to give up the illusion she had cherished all her life, that she should find her *real* sphere in the society she would meet in Paris; it might be among her own country people, but they would be broadened by travel and study until all desirable and agreeable qualities would be blended into a harmonious whole.

When she decided to pass the winter with her aunt, Mrs. Mongomery, it was with the sweet hope that she

should be able to realise her dreams
of a little "Salon"—a revival of that
delighful French institution and formu-
lated on the same lines, but having
American cleverness and adaptability
added to it. It seemed feasible. Mrs.
Montgomery had lived in Paris for
years, and she knew all the resident
society people, the rest of the "floating
population" were usually provided
with letters of introduction to her.
Her "Tuesdays At Home" were delightful functions. Katherine Cameron
had great respect for her aunt's discrimination, which often seemed pro-
phetic, and caused uninitiated people
to wonder *how* Mrs. Montgomery
happened to have "taken up" some
artist or singer who afterwards became
famous.

Still Katherine was not entirely
satisfied. Men liked her, but thought
her cold; at any rate, she never
fulfilled any promise of a flirtation that

her agreeable manners might suggest.
Women said she was ambitious, that
she would only marry some distinguished
foreigner, and yet Miss Cameron, who
sometimes used forcible expressions,
had been heard to say, "She would
marry a 'Hottentot' if she loved him."
She was honestly trying to get some
good out of her surroundings, and was
perfectly willing to fall in love, or to
gratify her intellectual tastes, just as
it might happen. Up to this time,
however, she had been distinctly
heart-whole, and aside from an occasional charming man or woman whom
she met in society, or the interesting
art students whom she knew (and liked
best of all), it seemed to her clear and
practical mind that there was a great
deal of "padding," as she expressed
it.

She resented, as a patriotic American
woman of culture and refinement, that
the so-called "exclusive" circles in the

American quarter accepted some of
the families who would not occupy
conspicuous positions in their own free
and enlightened country. She could
not help comparing certain wealthy
young society women with a clever
but poor friend of hers, whose artistic
talent had been recognized by her own
warm-hearted Southern townspeople,
who had contributed a sufficient sum
to send Miss Paterson abroad, confident
that her brush would one day repay
them. The two young women had
met at the studio of a common friend,
and Miss Cameron, who professed to
know nothing of art, had asked such
intelligent questions of the young
student that Miss Paterson, with a
woman's quick intuition, had surmised
that her fashionable countrywoman had
a more artistic nature than she admitted.
A friendship was begun, and Katherine
Cameron became the *confidante* and
admirer of the rising young artist.

Just now she has returned from a musicale at the hotel of one of the famous teachers, and she is describing it to Miss Paterson, who has come in for a chat and a quiet cup of tea.

"It makes me so indignant," she is saying, "when I think what an impression we must make on intelligent French people. Why this afternoon, at Madame de la Harpe's, it was simply one medley of disputing mothers and jealous pupils. Madame herself is so distinctly a lady, that when two irate mothers appealed to her as to which of their daughters should sing *first*, she shrugged her shoulders in true French fashion and said, 'They will both sing many times; they will sing so well that it will be doubtless required'—a diplomatic answer! She knew her audience, and felt that a programme of twenty-three numbers could not admit of many encores in one afternoon. I noticed she did not deviate from the original plan. Then that vulgar Mrs. Booth, from somewhere out west, who has the gorgeous apartment, and the family of extremely pretty daughters, asked me if I would join their French class. 'We have an actor, M. de Valle, to teach us,' she said, 'he is just splendid—so handsome and so polite; only he will make us *congregate* verbs.' To my horror, Mr. Vincent, of the English Embassy, who is so coldly critical of everything American, overheard her, and I saw him trying to suppress a smile, which made me indignant, so I impulsively replied, 'I shall be charmed, Mrs. Booth—so kind of you to ask me.' And now I shall have to extricate myself from that situation, for, although I have a certain appreciation of the ludicrous, I cannot sacrifice one night of every week, even to show Mr. Vincent that I despise his criticism."

"But I have rather thought Mr.

Vincent one of your admirers," Miss
Paterson returns.

"Admirer? He sees in me a young
person who will not be apt to make
any very ridiculous blunders, and as
he *has* to appear occasionally, being
in the diplomatic service, he talks to
me as a sort of compromise between
the tourist element and his own fixed
aristocracy. I *love* to shock him.
Why, to-day, he said, in that deliberate
tone he employs when he wishes to be
particularly patronizing, 'I suppose
you go in for all sorts of things, Miss
Cameron. I hear you are artistic, and
know the Latin Quarter better than
this side of the river. When do you
get it all in?' I told him to behold
a young person positively unique in
Paris—one who was actively pursuing
nothing. And then he actually remarked that 'in an age where all the
young women were running mad with
fads it was refreshing to find one so

confessedly idle.' He aggravates me so that I always lose my head, and get the worst of the argument. But here I am talking away, and forgetting that I am to hear all about you and your plans."

Miss Cameron soon proved that she could listen as well as talk, for she was most attentive while Miss Paterson told her about a letter which she had received that day, and which had disturbed her not a little. In the midst of their displeasure both girls saw the ludicrous side of it, for it was nothing less than a letter from Miss Paterson's townspeople *forbidding* her marriage to the penniless young sculptor with whom she had fallen in love.

"What impertinence!" Miss Cameron remarks; "talk about the tyranny of European courts! Here you are, an orphan, without a relative in the world to restrain you, and these people fancy they *own* you, and can control your liberty just because they have furnished you with funds which they ought to know will be returned to them."

"But there *is* a moral obligation," Miss Paterson replied. "I shall send them back every penny of their money as soon as possible, and I shall always feel a debt of gratitude which no pecuniary remuneration can cover."

"Little saint!" Miss Cameron exclaims, but she respects her brave little countrywoman all the more, because she is so visibly distressed at the situation.

"Let us go over the facts," adds Miss Cameron. "Here they are briefly: A number of your townspeople, seeing in you evidences of talent, raised a sum of money and sent you to Paris two years ago. Two of these people selected your masters (fortunately they made no mistake there); you have worked faithfully and conscientiously, and have accomplished more than most art students do in twice the time. This year two of your studies have been in the Salon, one of them was bought by a Frenchman of critical taste; and you have a number of charming saleable studies, besides your large picture of the garden-party intended for next year's Salon, in which festive scene your humble servant poses as the hostess serving tea to a group of *fin-de-siècle* society people. You are sure to make a hit with that, so many of the figures are actual portraits, and Paris dotes on personalities. It is conceded that merit no longer wins, but to be 'received' one must be a friend of some member of the jury, or paint the people whose vanity moves them to pull some wire, so that they may gaze down from the Salon walls upon an inquisitive and envious public."

"And in this case can I count on *you* or some of your admirers to pull the wires, Katherine?" Miss Paterson mischievously asks.

"Yes; that picture shall hang 'on the line,' even if I have to lobby for it; but you know all the artists think it splendidly treated," said Miss Cameron.

"I hoped it would be received this year, but, do you know, I have been considering all day whether I had better not sell it now, and send back as much money as I can raise immediately; for I intend to marry Edgar McDowald, with the benediction of my patrons if possible—without it if necessary," emphatically declares Miss Paterson.

"And I shall aid and abet you, especially if you intend to show them that 'love laughs at locksmiths'—and creditors. But, seriously, why not have an art sale? I am off to a musicale at that extraordinary Mrs.

Smyth's (formerly spelt with an i), who
begins every Monday morning sending
letters, followed during the week by
three-cornered notes marked '*pressée*,'
in which she 'begges' her dear friend,
whoever it may be, to run in Saturday
afternoon, and casually remarks that
some 'celebrated musicien' will per-
form. The joke is they usually do,
and we all find ourselves there once or
twice a season. To-night the American
Minister has promised to be present,
and I shall profit by the occasion to
invite everyone to your studio next
week to see some charming studies
which will be for sale."

Miss Paterson knew Miss Cameron's
influence, and felt that she was quite
safe in letting her friend have her way;
so after talking over the details they
separated.

That evening Miss Cameron succeeded in quietly scattering the
information through the crowded

rooms that a very charming friend of hers, the Miss Paterson, who occasionally received with her, would have a little private art sale the following week. Among the attentive listeners was Mr. Vincent, who casually asked if Miss Paterson had finished her Salon picture which she had described to him.

"She has," Miss Cameron replied, and suddenly added, "And you know, Mr. Vincent, I cannot offer my friend money, nor would she sell me so important a picture as her large one, for she would think I did it to help her; now, I want to ask you, as the person she would think of as being the last one connected with me (here Mr. Vincent smiled a rather melancholy but affirmative smile), to buy two of her studies for me in some other name. I can easily dispose of them as presents, and she will never be the wiser."

" Miss Cameron's wishes are my commands. I will call on Miss Paterson before Wednesday, and on the day when the exhibition takes place, you can be sure that at least two pictures will be marked 'Sold.'"

" That will give a business-like air to the whole arrangement, Mr. Vincent, and suggest to any possible buyers that other equally attractive studies are for sale. This must be a profound secret. Do you promise? "

"Certainly," Mr. Vincent replied, and Miss Cameron knew she could trust him.

" He is really very likeable, when one sees him alone," Miss Cameron soliloquizes; and then she reflects that it is decidedly her fault that she does not see Mr. Vincent more frequently in his best light; she remembers various occasions when she has made their duet a trio by addressing some third person, thus preventing a possible tête-a-tête.

The afternoon selected by Miss Paterson arrived, and as Miss Cameron alighted from her coupé in the humble street where art and poor students hold sway, she remarked with pleasure a goodly line of private carriages, and knew that her scheme had succeeded, and that Miss Paterson was the fashion—at least of the hour. The question was, Would they buy her pictures? And then she added to herself, "They must be sold, even if I have to find other agents, and buy them all in." But the loyal girl might have spared herself any anxiety. As she entered the room, which was artistically draped and hung with numerous strongly-executed sketches, she saw the magic word "Sold," not only on several of the small studies, but conspicuously placed at the base of the largest canvas, Miss Paterson's salon picture, in which Miss Cameron is the central and principal figure.

"Isn't it too delightful, dear?" Miss Paterson whispers to her. "An Englishman, a friend of Mr. Vincent's, came here with him yesterday, saw my canvas, liked it, asked my price, and actually took it. Mr. Vincent also bought two other studies, and several have gone to-day. Edgar has lost no time. He has disappeared now to cable to my esteemed benefactors, '*Marriage will take place; cheque for full amount on way.*' Extravagant of us, I know, and of course it's extremely '*previous*,' but we really see our way clear to happiness, and I shall always feel *you* did it all."

As Miss Cameron shook hands with Mr. Vincent that day she told him that he had been instrumental in making two deserving people happy.

"It was so thoughtful to bring your friend here, who bought the large picture," she says. And then she adds, "Did I ever see him?"

"I think you have seen him," Mr. Vincent replies. Something in his man ner betrays him, and Miss Cameron, guessing the truth, impulsively says :

"You bought it yourself, Mr. Vincent."

"Hush!" he softly whispers, with his finger on his lips. "We are fellow-conspirators, and cannot betray each other."

Next year, when a great American city gave Edgar McDowald the order for a State monument, the beauty of his designs having distanced all competitors, Parisians remarked that Mrs. Montgomery's discrimination, as regarded celebrities, seemed to have fallen upon her niece.

Mr. and Mrs. McDowald delight in telling of their romantic courtship, and how Miss Cameron's scheme of an art sale brought about their marriage ; but Miss Cameron always affirms that its success was not due to her, but to

Mr. Vincent's tact in exhibiting that expensive canvas to his friend.

Miss Cameron, being a worldly-wise young woman, tries to feel that Mr. Vincent's motives were wholly generous and disinterested; but if what rumour says is true, Mr. Vincent would do more than that for the charming central figure in Mrs. McDowald's Salon picture, which now looks down from a good position in the library of his own English home, and which never hung "on the line" after all.

V

A Complex Question.

THERE were a half-dozen or more good riders in Tangier that winter, but Bob Travers was the acknowledged leader. At every annual race-meeting he proved to his backers that their confidence in him was not misplaced, for, brave fellows as they were, none of them rode so hard, or cared to take the risks which Bob cheerfully ran.

Robert MacNeil Travers, familiarly known as "Bob," was spending his second season in Africa. The first time he had run across from "Gib" to look up something in the way of horseflesh, and once there he had easily fallen in with a set of men whose society he enjoyed extremely. They were dashing fellows, several of them young. English noblemen, who found the free, bold life they could lead in this lawless place too fascinating to leave. It was very agreeable in that delicious winter climate to dash off over the wild country on a sure-footed Barb horse, or to join some caravan for a few weeks' excursion in the interior, while in England everyone was freezing, or at least imbedded in fog.

They had their little glimpses of civilization—the Tangerines—for the few resident Europeans were very glad to entertain any interesting visitors from the outside world. Bob Travers was as much liked by the wives and sisters of his friends as any gallant, well-bred Englishman deserves to be, and every one was pleased when his engagement was announced to pretty Mabel Burke, the sister of Boardman Burke, the artist, whose Eastern scenes, painted under the clear skies of Morocco, have won for him the reputation of being one of the foremost exponents in the new " Impressionist School."

The occasions were rare when Bob Travers was not included, whether it was for a boar hunt, a day with the fox hounds, or a little dance, at any one of the half dozen hospitable European houses.

One night he was late in arriving at a dinner-party given in honour of some Americans, whose yacht had appeared in Tangier Bay that day; they were already seated at the table when Bob slipped quietly in, and, at a little nod from Miss Burke, found his place beside her. He was concious that his other neighbour was a woman—a young and attractive one. He had time to observe that, when his obliging hostess, in reply to his apologies, said, " You are punished enough, for you have lost at least ten minutes of Miss Schuyler's society." This, with a knowing little look at Miss Burke, which seemed to say, " To be sure he is your property, but if you are engaged to the most present-able man in Tangier, you must pay the penalty, and give him up to occasional and fastidious visitors."

Modest little Mabel Burke, who simply basked in " Bob's " smiles, and wondered at her own good luck in ever winning his love, gave her hostess a proud, happy glance that spoke volumes for her sense of security.

A closer look at Miss Schuyler convinced Mr. Travers that he had never met anyone at all like her; she was so self-possessed and clever that they were soon talking as freely as if they had been old acquaintances. She was not so pretty as his *fiancée*, but she was very fascinating (a charm

that even Bob had not attributed to
Miss Burke), and her versatility amazed
him.   It did not seem to matter
whether they discussed horses, religion,
or politics—Miss Schuyler had her
opinions, and she expressed them
without conceit or aggressiveness.
During the fortnight that the smart
little yacht *Liberty* was anchored in
the waters of Tangier Bay, and its
merry party were devoting their days
to long country rides, excursions to
Cape Spartel, or cantering along the
sandy beach, Travers found Miss
Schuyler the most interesting of companions; he seemed to have become
her acknowledged escort, and (since
one night, when he had nearly killed
his best horse by galloping several
miles for a doctor to come to the
rescue of one of the ladies who had
broken her arm while the party were
making an excursion) Miss Schuyler
had singled him out for all sorts of

delicate favours. He, on the other hand, discovered that this woman, with her grace and culture, was just such a woman as he had pictured he should eventually take to Travers Towers as its mistress. For in less than a fortnight he realized that in his happy-go-lucky way he had drifted into that engagement with the pretty sister of his dearest friend. What could be more natural? All the conditions had favoured his courtship, and until he saw Miss Schuyler it had seemed very agreeable to possess the affections of the nicest girl in Tangier.

He knew she was not the wife he had dreamt of, but then, he reasoned, one never marries one's ideal. Mabel Burke was sweet and good, and loved him; so one delicious, star-lit night, after a cosy dinner, he found himself alone with her in the quiet little Moorish court of the Burkes' villa, and as Mabel gave him his second cup of coffee he looked at her approvingly, and on the impulse of the moment told her he should like to have her always with him. He meant it then; and after that it was all easy sailing, for Boardman Burke was delighted to give his sister to a man whom he already loved as a brother. The gossip of the town had not reached the visitors in the yacht, and Miss Schuyler only heard accidentally that Mr. Travers was engaged to Miss Burke, for Bob had felt a reluctance to tell her—had supposed someone else would—and, finally, seeing she believed him to be free, he had *dreaded* to tell her. And so their relations progressed undisturbed, and, like all things under an Oriental sun, developed rapidly.

They had been taking tea at Mr. Boardman Burke's and looking at his pictures, when suddenly the artist said:

"I must show you the one I am doing for Travers' wedding present."

And when someone remarked that he could take his time to finish the painting, Boardman Burke had said very distinctly:

"Oh, no! I expect to have to give my sister, as well as that best picture of mine, to Travers before the year is out."

It is just possible that Mr. Burke thought it wise to make this statement, for occupied though he was in his work, he had observed that his sister looked troubled. Although Travers dropped in every day, he, too, seemed pre-occupied, or was in a hurry, and he was seen constantly riding with Miss Schuyler. Little Mabel was too seriously in love with him, and believed in him too deeply, to admit that he had been the least remiss in his attentions to her, but she felt relieved, all the same, to hear that the *Liberty* would hoist anchor and go over to Gibraltar the next morning, and from there continue her course along the coast of Spain and the Riviera. Even when she heard Travers and the American Consul accept an invitation to go to Gibraltar with the party, she felt no uneasiness, for he would return the following noon by the regular steamer. So she let her accepted lover stroll off with Miss Schuyler, only saying a quiet "good-bye."

When she looked out from her window the next morning the pretty little yacht had disappeared, and all day she fancied Bob buying up supplies, which he said he wanted for an expedition into the interior.

In reality, when Mr. Travers had glanced at Miss Schuyler, after the announcement made by Mr. Burke of his engagement, he thought she looked a trifle pale, but then there is such a peculiar light when the African sun comes down into a Moorish garden through the waving palms that one gets strange impressions.

Miss Schuyler was very silent on her way to the beach, and Travers did not see her again till morning, when he crossed on the yacht to Gibraltar. During the night a sense of all he had lost flashed upon him; he could see no way out of it. He was a man who prided himself upon keeping his word; that word was given to Miss Burke, whom he liked and respected, but whom he now knew he did not love. And he had allowed himself to drift on through two happy weeks, devoting himself to this stranger, who in return must certainly despise him for his cowardice. Distinctly, it was an awkward position. He felt confident that, given his freedom, he might win the woman of his choice, for she was the kind of woman to inspire him to

do his best, and Bob Travers' best
was very good indeed, but his freedom
was just what he could not ask for, so
he finally decided to tell Miss Schuyler
the exact truth, and thus at least feel
he had her respect.

On the yacht he told her his story,
and she listened, as a woman listens
who has had many disillusionments,
and accepts them as necessities.

He thought her very cold when she
only said :

"We have been very good friends,
Mr. Travers. It will be enough
to tell you first that I should have
preferred to hear of your plans from
your own lips. It all seemed so
natural in Tangier, so far from the
conventional outside world, that I
allowed myself to give way to impulses
which I thought under perfect dis-
cipline."

"But you must know, you *shall*
know, that my heart is yours, that you

are my *ideal* woman, the one I should have married," Travers earnestly pleaded.

. "If that is so, let it encourage you to be strong. Go back, marry your little girl, and forget one who has suffered too much to judge anyone." Then Travers went down the side of the yacht into a small boat, and could only say "God bless you" over her extended hand before the steps were pulled up, and the yacht steamed out on her way to Malaga.

A few days after at Marseilles the papers were brought on board, and an article in them instantly attracted their attention. It graphically described a fatal accident that had befallen Robert MacNeil Travers, who had just landed from a yacht at Gibraltar evidently in perfect health. He had gone up to the summit of the rock, and stood at the edge of its dangerous eastern face. His companion, the

American Consul at Tangier, had
stopped a moment to look out to sea
with his glass, and when he turned
round poor Travers had disappeared,
" probably seized with vertigo," the
paper said ; for Mr. Travers was heir
to a large estage, and about to be
married to the sister of the celebrated
artist, Boardman Burke, so no idea of
suicide was entertained.

Who shall say whether Miss Schuyler believed this newspaper version?
Perhaps she remembered Travers' last
impassioned word, " You *shall* know
my heart is yours," and he had taken
this way, the only possible way, to
show her his devotion without being
dishonourable.

*   *   *   *   *

Poor little Mabel Burke wept griev-
ously, but she is again engaged, this
time to a man who is far more domestic
than poor Travers.

And Miss Schuyler? She continues to be Miss Schuyler, although she is as fascinating as ever. A woman who has tested one man's affection to the death and not found it wanting, is not easily won !